Dieter Mende
EEZ Energy, Energy industry, Future energies

The siren-like songs
and the fairy tales
of full electrification.

The sustainable energy transition links
the electricity sector with heat,
gas products and fuels.
Hydrogen is the link with Power-to-X.

Resisting the siren songs,
the energy transition with hydrogen,
the challenges and their implementation,
The energy lobby versus the energy transition,
the prospects.

If someone tells you that they can explain the energy transition to you within a few minutes, then you should be extremely skeptical.

Thinking at EU/federal/state level and acting locally is not a contradiction, but rather a dynamic energy policy.

Dieter Mende

Dieter Mende
EEZ Energy, Energy industry, Future energies

The siren-like songs
and the fairy tales
of full electrification.

The sustainable energy transition links
the electricity sector with heat,
gas products and fuels.
Hydrogen is the link with Power-to-X.

Resisting the siren songs,
the energy transition with hydrogen,
the challenges and their implementation,
The energy lobby versus the energy transition,
the prospects.

Impressum

© 2024 **Dieter Mende**, EEZ Energie Energiewirtschaft Zukunftsenergien
www.eez-mende.de

weitere Mitwirkende:

Antje Mende,   LIKES Layout – Impuls – Konzept – Entwurf – Style;
Moderne Medien-, Text- und Bildberatung

Production and publishing: BoD – Books on Demand, Norderstedt

ISBN: 978-3-7597-0664-5

Inhaltsverzeichnis

Prolog

Understanding the energy transition.

The energy transition is much more than just the increasing use of renewable energies!
The energy transition is also a booming global job engine.

Picture:
Copper plate EU; even if the electrical grid were to be 100 times larger than it is today, only as much electrical energy could be fed into the grid as is taken from the grid elsewhere at the same time to protect against grid overload. Without storage, we remain with the curtailment of the generating plants, such as the wind turbines, which means that electrical energy that can already be generated today is lost, which means that we will continue to have higher costs for electrical energy in the future, because the operators of the generating plants, such as the wind turbines, are entitled to financial compensation for the curtailment. Please do not let yourself be unsettled; Dieter Mende, EEZ Energy, Energy industry, Future energies.

The success of the energy transition therefore depends crucially on the start and speed of implementation of the defined goals, which requires determination and regional identity with the fields of action.

There is an old saying:
„Sitting on the shoulders of a giant makes it easy to see new horizons."

The Emscher-Lippe region in the northern Ruhr area of Germany has also been hit hard by the closure of the mines and the resulting structural change.
The Emscher-Lippe region has not been able to call for a giant to shoulder the region and carry it forward to new destinations; the challenges for the region are still numerous today.
The regional nucleus h2herten with its expansion into supra-regional commitments with the h2-netzwerk-ruhr is exemplary.

If the energy transition succeeds in the demanding and densely populated Emscher Lippe region, then the energy transition can succeed everywhere!
Climate change is already clearly visible and the consequences are coming faster than feared.

One of the three words of the year 2019 in Europe is climate youth!

If the energy transition is to succeed, there is no alternative to the expansion of renewable energies. If the targets set with regard to phasing out coal are to succeed, politicians must ensure that the conflicting political decisions are changed immediately to ensure the success of the energy transition. These current inconsistencies are the subject of warnings from people in the EU.

People in the EU have been aware of the fact that world peace is also directly dependent on the success of the energy transition, and not just since the latest climate developments. The effects of climate change hit people in poorer countries particularly hard, as well as people in Africa.

If the living conditions of the poorest continue to deteriorate and people move to the more temperate zones of the world, this will have a significant impact on world peace.

People in the EU have long been unable to understand and now no longer accept the contradictory statements on the climate reports by those who are lobbying in favor of fossil fuels; the slowing down of the energy transition has already had a major impact on the outcome of the 2021 federal elections and will increasingly have an influence on the outcome of the elections in the EU.

Those who are involved in well-paid lobbying for the current coal policy with deliberately incomplete impulses, with the aim of unsettling the population, are now faced with the decision as to whether their actions are justifiable, whether greed for wealth is weighted higher by lobbying than responsibility for future generations, at the latest in view of the current climate reports.

The fact that insurance companies are also investing in the technologies and infrastructure of the energy transition is justified by the need to avoid increasing losses, such as those caused by more extreme weather due to climate change.

Citizens' wind cooperatives are an example of how the energy transition in conjunction with citizen participation not only has a positive effect on reducing climate change, but also offers a very interesting investment opportunity for people, in which everyone can decide for themselves how much money they want to contribute to citizen participation. The energy transition has become a global job engine, also thanks to the technology expertise, infrastructure expertise, energy expertise and storage expertise from Europe.

Antonio Guterres (UN Secretary-General) was prompted to call on politicians to act quickly in view of the results of the World Climate Report of August 2021.

Svenja Schulze (Federal Environment Minister; SPD) warns with regard to the results of the World Climate Report of August 2021 with the words:
"The planet is in mortal danger".

„Resisting the siren songs of full electrification"

This book has a broad target group:
+ Citizens interested in the energy transition,
+ Students and teachers,
+ Decision-makers in the cities,
+ Decision-makers in politics,
+ Entrepreneurs seeking orientation and inspiration,
+ Product developer,
+ Engineers,
+ Entrepreneurs,
+ Founder.

Resisting the siren songs and fairy tales of full electrification

The name Odysseus is known from Greek mythology. Odysseus was given the rule of Ithaca by his father. In the course of one of his journeys, Odysseus passed the island of the Sirens and was aware of the danger that any man who succumbed to the lure of the Sirens' sweet song and headed for their island would be lost and die. Aware of this danger, Odysseus glued his traveling companions' ears with wax and had himself tied to the ship's mast, so that Odysseus and his traveling companions could safely complete their journey home to Ithaca.

With regard to the challenges of the energy transition, there are also tempting siren songs from the lobby, which often want to lure people into their market positions with deliberately incomplete contributions.

People should certainly not close their ears to the energy transition.
However, attention is required when promises are formulated with trivial-sounding answers to the many complex tangents of the energy transition.

A great deal of attention is required when quick successes are predicted for the many complex tangents of the energy transition.

This energy transition is not the first energy transition.

With the first energy transition, from burning wood to burning coal, fears were created and it was claimed that Europe would face economic ruin with this transition because it would allegedly not be possible to extract so much coal from the ground for static reasons so that there would be enough fuel in the EU. It was also claimed that the massive extraction of coal could even have a major impact on the tectonics of the European tectonic plate, including the most severe earthquakes or even volcanism in Europe. The deliberate false reports were the result of lobbying by the timber industry.

With the second energy transition, from burning coal to burning natural gas, fears were again created and it was again claimed that Europe would face economic ruin with this transition, because for technical reasons it was supposedly not possible to supply such a large volume of gas over such a long distance at the same pressure so that there would be enough fuel in the EU. The deliberate false reports were the result of lobbying by the coal industry.

With the current energy transition, the pro-fossil fuel lobby is once again creating fears and claiming that this transition would spell economic doom for Europe. Allegedly, power cuts would be the inevitable consequence.

The energy transition is not trivial!

As the pro-fossil fuel lobby can no longer hold its own in the complete energy transition considerations, from energy generation to energy applications, the pro-fossil fuel lobby has moved on to distorting the energy transition with the aim of creating uncertainty through deliberately incomplete, even deliberately false statements, with the juxtaposition of potentials that actually complement each other.

Those who only look at the physics of efficiency cannot really contribute to the global challenges of the energy transition. Those who only allow the physics of efficiency in the overall view must ask themselves why people drive a car with an internal combustion engine, even though only 27% of the energy originally used reaches the tires on the road in terms of efficiency.

Opponents of the energy transition stubbornly ignore the fact that renewable energy generation in the terawatt range is curtailed in order to protect the grids from overload; this is electrical energy that can be generated in principle and is irrevocably lost when the generating plants are curtailed. Storing the electrical energy that can be generated is therefore important and correct.

Storing electrical energy as an alternative to shutting down the generating plants is also an option for hydrogen, but the opponents of the energy transition are using physics to get in the way of efficiency.

Not only from a physical point of view, but also from an economic point of view, with the aim of reducing energy costs, hydrogen production is a value-added alternative to curtailment.

This is because the shutdown means:
+   pro rata system costs despite the curtailment,
+   Compensation payments to the operators for the
        curtailment of the plants,
=   No power generation.

On the other hand, hydrogen production:
+   the pro rata investment costs are constant,
+   NO compensation payments to the operators for the
        curtailment of the plants,
=   However, the cost reduction of energy generation
        through supplementary value creation.

Opponents of the energy transition aim to create uncertainty by placing the storage of electrical energy in competition with hydrogen, whereby the future potential is claimed with a view to the development of battery technologies, but the future potential of hydrogen technologies is not conceded.

The results of the studies on the issues of the energy transition, with a holistic view, show that the coupling of the electrical energy sectors with gas products, with fuel products and with heat triggers the maximum technical potential.

The results of the studies on the issues of the energy transition, with a holistic view, also show that the energy industry is not being turned upside down at all, but rather that the very many expansion potentials optimize the desired results of the energy transition.

If, contrary to the results with regard to sector coupling, there is still no uniform understanding of the path to the energy transition, then the reasons for this can be found in the clearly incomplete lobbying of the individual energy paths.

The idea of the consequences of full electrification is deliberately not communicated, because if the energy industry wants to generate heat with electricity, the additional demand for electricity would be enormous.

Moreover, this is in a situation in which we have to realize that we are far from being able to solve the challenges of phasing out nuclear and coal energy with the plants for generating electricity from renewable sources that have been announced so far.

The siren call for full electrification lures people in with the simple and convenient-sounding claim that we only need to be able to provide enough renewable electricity and all energy issues will be solved, including mobility.

However, even if the electrical grid is expanded up to a hundredfold, the systems generating the electricity would still have to be regulated to protect the grids from overload, because only as much electrical energy can be fed into the electrical grid as is being drawn from the grid elsewhere at the same time.

The expansion of renewable electricity generation makes it possible to reduce CO2 emissions and achieve the goal of reducing the effects of climate change.

This sounds simple and coherent at first, but what is being concealed is:
+   the enormous costs for the massive grid expansion,
+   the fact that electrical storage is also necessary
        to reduce costs,
+   because every single kilowatt hour of electrical
        energy generated on site is a valuable contribution
        to reducing energy costs.

The term "copper plate Europe" was coined with a view to the resulting scenario, and only with a view to the massive expansion of the grid; a term that represents the foreseeable enormously high costs.

The enormous amounts of energy in the terawatt range, which are to be generated with the planned wind turbines, and which technical solutions will then be used to store them, are being concealed.

In the gigawatt to terawatt range, we really no longer need to talk about storage with batteries alone.

Batteries and hydrogen complement each other:
+	local energy storage with batteries,
+	regional energy storage with batteries and hydrogen,
+	neighborhood solutions, e.g. apartment blocks or commercial areas, with batteries and also with hydrogen,
+	Supra-regional energy storage with hydrogen,
+	national energy storage with hydrogen,
+	international energy storage with hydrogen,
+	international energy transportation with hydrogen.

Since the energy transition is also a move away from material-based energy sources (energy bound up in coal, oil and natural gas) towards renewably generated energies that are electricity-based, hydrogen is not only a classic energy storage medium, but also an energy source that acts as a connecting element for sector coupling and triggers a variety of technical possibilities, which have been given the collective term Power-to-X.

Power-to-X includes, for example, the potential for:

+   Power-to-Gas,
+   Power-to-Heat,
+   Power-to-Liquides,
+   Power-to-Chemicals,
+   Power-to-Fuels.

Farmers are increasingly also becoming energy farmers.
With the increasing problems caused by climate change,
Power-to-X also has the potential to:
+   Power-to-Agrar.

Globalization has expanded online retail.
With the increasing problems caused by climate change,
Power-to-X also has the potential to:
+   Power-to-LML Last Mile Logistik.

Climate change and the necessary drinking water.
Power-to-X involves the increasing problems:
+   Power-to-RDW recovery of drinking-water.

Pictures:
Hydrogen user center h2herten with the hydrogen filling station;
Dieter Mende, EEZ Energy, Energy industry, Future energies

Picture:
The "white tower", the pressure accumulator at the H2 user center h2herten;
Dieter Mende, EEZ Energy, Energy industry, Future energies

Who wants to invest in a future with a great deal of uncertainty?
Who wants to invest in the energy transition?

Which part of the population in the EU should want to invest in the implementation of the energy transition if deliberately unsettling statements are spread about the aforementioned drivers for implementing the energy transition?

The drivers for implementing the energy transition are known as:
+ the expansion of renewable energy generation,
+ the simultaneous reduction of fossil fuels.
+ Energy grids are becoming flexible with the sector coupling of electricity, heat, gas products and fuel products,
+ through the energy carrier coupling the sectors Hydrogen.
+ The increase in energy efficiency,
+ the simultaneous development of components and devices with lower energy consumption.

In view of these points, it is very surprising why these promising drivers of a sustainable infrastructure in the energy transition have not yet led to implementation.

It was not only the Fridays-For-Future protest movement that identified the then Federal Minister of Economics, Peter Altmaier, as the driving force behind a tough lobbying policy.

Is the European economy being overtaken by the energy transition as a job engine? It wouldn't be the first time.

Hybrid technology was developed in Germany, was (deliberately?) completely misjudged by German industry and was only integrated into the systems after the markets in Asia led the way with hybrid technology.

Only when the statements on the energy transition become visible to the people, with their implementation by the industry and the economy, implementations that make it possible to achieve the climate targets, can people's trust follow.

The market ramp-up of future energies, including hydrogen, will be successful with recognition in the holistic view, flanked by proactive energy transition marketing. The importance of proactive marketing is demonstrated by natural gas vehicles.

The marketing for natural gas vehicles has not just been bad, there has been no recognizable marketing for natural gas vehicles at all. As a bridging technology, natural gas vehicles were supposed to be the first step towards energy efficiency, developed by the automotive industry to solidify the technology of combustion engines in the markets.

Explaining the intentions and goals was neglected, so that the intention behind them could not be understood.
At the same time, the justified doubts about the concept of natural gas vehicles were never answered by the automotive industry, so that no acceptance could develop.

Never before, and not only in Europe, have market trends emerged solely as a result of technical developments. With regard to successful markets, there has always been an initial formation of acceptance; be it out of a mainstream, be it out of necessity.
Without the creation of acceptance through suitable marketing, the potential has never emerged from the niche markets.

With regard to the energy transition, the opposite is true. There is a very high level of acceptance here, not only in Europe; in this case triggered by environmental necessity. In this case, it is industry and business that have so far failed to make consistent use of the opportunities.

However, it has always been the case that a successful industry and economy has integrated future opportunities into the current fields of the markets in good time.

It is therefore not just a lack of consistent guidelines on the part of politicians, as is often criticized by industry and business, that has prevented the necessary steps in the energy transition from being implemented.

These contradictory statements on the part of industry and business as well as politicians have been noted by environmental associations for many years; however, industry and business have not responded with an explanatory position that would have made it possible to work together to achieve the climate targets. Industry and business have responded to the statements made by environmental associations by commissioning studies which, in their initial formulation, often had the purpose of creating uncertainty among the public with the results of the studies and their incomplete consideration of the interrelationships.

In view of these recognizable correlations, which do not provide the necessary framework conditions for the energy transition, it becomes clear that in the course of the election campaign for the 2021 federal election in Germany, people no longer wanted to follow the statements of those who were politically responsible for the energy transition in 2021. The Ministry of Economic Affairs, led by the CDU, had clearly acted against the goals formulated by the Ministry of the Environment, led by the SPD,
Yes, the energy transition is not trivial, people in Europe have realized that. All the more reason for many people to take a very close look at who they entrust with the urgent implementation of the climate targets. In view of the effects of climate change and environmental problems, there is no justifiable reason for the people of Europe to continue to delay the implementation of the energy transition.

The Stone Age didn't end because there were no more stones, and the hydrogen industry won't only get started when there is no more oil.

The conversion of the energy supply to a hydrogen energy economy is often said to take around 50 years; the conversion from coal to oil with natural gas is used as a benchmark. However, it should not be overlooked that when the industry talks about a transition, the following situation is assumed:
+ security of supply at all times at any place at any time
+ A nationwide infrastructure:
  + the supplier has sufficient distribution options available,
  + the customer can choose between the products at any products at any location,
+ an economical energy price.

This is probably the all-important aspect of our dependence on fossil fuels:
At what point, taking into account various consumption trends, the production of fossil fuels will no longer be able to meet energy demand.

Energy demand in Europe is often underestimated when it comes to heat; in the EU, more than 50% of energy consumption is used to generate heat.
Hydrogen as an energy source can also offer technical solutions for heat with power-to-heat; probably not in every house, as is the case with natural gas, but in neighborhood solutions.

A sustainable energy supply with the important security of supply relies on new technology options being ready for the market at an early stage and being both economical and environmentally friendly, with the aim of achieving climate neutrality.

Following the increased closure of mines, interest in mine gas as a possible additional source of energy has grown steadily. Even after the end of mining, more than 120 million standard cubic meters (Nm^3) of mine gas escapes unused into the atmosphere from old mines every year.
The resulting drop in gas pressure in turn leads to the release of adsorptively bound gas.
This effect is comparable to a champagne bottle, where once the cork is removed, the carbon dioxide bubbles until there is none left.
In Germany, 1.5 to 1.7 billion Nm^3 of mine gas are released in mines every year.

The methane content (CH_4) of mine gas is approx.
70%. The comparison of climate-relevant emissions in Europe makes it clear that the use of mine gas can also become important as a bridging technology, as storage for hydrogen.

The federal statistics on climate-relevant emissions in Germany per year are exemplary:
+   Mining: 568.359 [t/a] CH_4
+   Livestock farming: 22.300 [t/a] CH_4
+   Landfill: 183.295 [t/a] CH_4
+   Wastewater treatment: 107.280 [t/a] CH_4

Regardless of whether the shutdown of power plants is politically motivated or whether the shutdown of power plants is due to ageing, the existing infrastructure with a view to the grids and buildings has a great deal of future potential.
A real opportunity for fuel cells in power plants can be found in fuel cell types with a very high operating temperature (hot modules), as this means that local heating customers can continue to be supplied.

The four largest Japanese automotive companies alone are investing 700 million euros of their own funds per year in the series production of fuel cell drives.

This closes a very important circle; we come to the topic of human capital.

Human capital

The skilled workers required for the production companies are available on the labor market in Europe due to the sharp decline in the heavy coal industry.
The current challenge facing the economy is to provide suitable training places and the appropriate retraining and further training of employees.

The success of German industry in international competition depends crucially on the start and speed of the transformation processes in all company structures, which requires the willingness to cooperate and the transformation of in-company training.

Just twenty years ago, skepticism and the hesitant attitude of industry towards hydrogen technologies were attributed to a lack of willingness to take risks or innovate. At best, environmental protection and the desire to establish their own technologies in the potential grid of the incipient discussion on the energy transition prompted companies to participate in the first pilot projects.
Today, competition in the potential grid of the energy transition is no longer just a strategy; the aim is to secure the future viability of companies with regard to their products and services.
Missed opportunities in the transition period could have shortened the path to the economy with hydrogen as an energy carrier!

Here is a significant example from Germany:

Investigations by the Hamburg electricity company HEW in 2007 showed that one million cars with fuel cell drive could be operated in Germany with the hydrogen that can be obtained from primary and secondary power plant control; this is based on the assumption that the cars have a fuel consumption of 10l/100km and that the cars have an annual mileage of approx. 12,500 km. By integrating the city buses running on regular routes (approx. 7,000 buses in Germany), another 3/4 million cars could be powered by hydrogen. This would be possible with practically no significant increase in emissions from energy production.

In the past, market acceptance has been decisive for the breakthrough of new technologies, and this is still the case today.
Today, product quality is once again winning out:
+ provided the price is reasonable for a private family household,
+ provided that the operation is relatively economical.

The initially hesitant market entry of cell phones and notebooks followed a similar course to the market entry of fuel cells. With the integration of the technical possibilities offered by hydrogen energy storage and fuel cells as energy converters, the existing value chains are supplemented and new value chains are opened up.
It is the value chains that define the fields of action and not the other way around; this also applies to the holistic view of the potential of the energy transition.

If someone tells you that they can explain the energy transition to you within a few minutes, then you should be extremely skeptical.
Thinking at EU/federal/state level and acting locally is not a contradiction, but rather a dynamic energy policy.

Picture:
The energy transition as a job engine: the technological expertise and infrastructure know-how that is emerging from the energy transition is in demand in Europe and globally.
Dieter Mende, EEZ Energy, Energy industry, Future energies
Picture consulting by Antje Mende, LIKES Layout Impuls Konzept Entwurf Style

Bridging technologies and future technologies

"Building bridges for the future" is an industrial incentive that began with the start of industrialization. The current energy transition is not the first energy transition in the world:
+	The burning of wood has been followed by the burning of coal.
+	The burning of coal was followed by the burning of oil and natural gas.
+	The combustion of coal, oil and natural gas has been expanded to include nuclear energy.
+	The imminent phase-out of coal and nuclear energy with the increasing reduction in the combustion of oil and natural gas will be followed by the increasing expansion of renewable energy generation, e.g. with wind, solar, biogas and geothermal energy.
+	Since the current energy transition is also the path from material-based energies, energies tied up in coal, oil and natural gas, to regeneratively generated electricity, special attention is being paid to energy storage in the giga-watt range with hydrogen.
+	Hydrogen is very well suited for storing electricity generated from renewable sources, with the intention that electricity generated from renewable sources is primarily used directly, and that electricity generated from renewable sources that our grids cannot absorb is converted into hydrogen for storage.

With the appearance of "biofuel" at filling stations, quite a few farmers have applied to the EU for a change of use of their arable land and fields.
The EU had to counteract this, so that the industry requested the EU to make a decision as to whether the industry should continue to focus on energy products in the direction of "biofuel" or whether the industry should start to focus on energy products in the direction of hydrogen.

For a long time, the lobby of the combustion engine and motor vehicle industries had been promoting the continuation of efforts in the direction of "biofuel"; currently, the lobby of the combustion engine and motor vehicle industries is promoting e-fuels, knowing full well that there is not enough potential for this, with the main aim of slowing down the energy transition for as long as possible so that the current business areas can experience the highest possible profit maximization.

The energy transition is much more than the increasing use of renewable energies.

The fact that the energy transition is an industrial policy platform with clearly formulated demands from both the EU and the federal and state ministries is evident in all media. In the EU, energy supply is initially based on security of supply, cost efficiency and environmental compatibility.

Back in 2008, EU Energy Commissioner Andris Piebalgs said: "... it is time for a European energy policy ... the completion of the internal market, climate protection and security of supply are the major challenges in the energy sector that require common solutions."

Since the energy transition is also the path from today's material-based energies, tied to coal, oil and natural gas, to an electricity industry with increasingly renewable electricity, energy storage systems are undoubtedly a necessary component of sustainable energy infrastructures, for energy supply security, for energy transport and for the provision of energy in the private, commercial and industrial sectors.

We take Germany as an example because Germany is a driving force in the EU with a view to the opportunities and possibilities offered by the energy transition. NRW North Rhine-Westphalia is the number one energy state in Germany.

Reports in the media give the initial impression that Germany, especially the area around Hamburg and Stuttgart, knows how the energy transition works.
But why don't the innovative regions in NRW receive more attention with regard to the energy transition? The fact that many people in NRW have problems with the description of the energy transition has a very different background.

On the one hand, the numerous brochures and flyers at the trade fair stands are often very specifically tailored for quite a few people and do not complement each other; on the other hand, some of the brochures and flyers are rather confusing and the brochures and flyers are therefore more likely to unsettle people than to show the energy transition in North Rhine-Westphalia in context.
Secondly, the energy transition in North Rhine-Westphalia is not immediately apparent as a holistic process because the individual core competencies are already clearly evident regionally and in the municipalities, but because their networks, cooperations and synergies are often unknown.

Even the efforts of people interested in the energy transition to find information via the usual search engines on the Internet, e.g. Google, often generate uncertainty because the searchers often have to search very specifically for terms, which in turn requires a great deal of prior knowledge.

The reason why the area around Hamburg and the area around Stuttgart often appear in the media lies in the fact that:
+ Hamburg communicates with one voice as both a city and a state,
+ Stuttgart communicates as a regional nucleus with one voice,
+ there are many voices in the numerous regions of NRW, which do not necessarily complement each other.

In North Rhine-Westphalia, for example, people are aware of the regional competencies, as the examples:
+	E-mobile city of Dortmund,
+	climate/plant/city/Essen,
+	Solar City Gelsenkirchen,
+	Innovation City Bottrop,
+	Hydrogen City Herten,
+	Fuel Cell City Duisburg,
+	Chemical City Marl,
+	Environmental City Gladbeck,
+	and many more.

However, the holistic Ruhr metropolis, for example, is not recognized in the potential grid of the state of North Rhine-Westphalia on a par with Hamburg and the area around Stuttgart. Cooperations, such as the cooperation between the h2-network-ruhr and HyCologne, are only known to those who have already dealt more intensively with the energy transition and energy storage systems.

A sustainable infrastructure in the energy transition is complex, full of opportunities and also very exciting at the same time.

With the innovation of the German energy industry, the conventional energy sources of crude oil and natural gas have established themselves in the energy markets alongside the coal industry since 1951. The decisive role of an energy location, an energy region, is probably its ability to generate and supply energy economically, efficiently, ecologically and sustainably; at affordable prices for the population.

Both an excellent environment with universities and research institutions, as well as economic and technical expertise, are the innovative basis; with regional economic development, the innovative engine.
This was true for conventional energies in the past and is also true for renewable energies today.

If the value chain of green electricity generation is closed by suitable energy storage systems, these additional energy yields will, from the point of view of the EU and the state and federal ministries, ensure that energy remains affordable for people in the future. It is already clear today that the EU's foresight is justified. Technological know-how and infrastructure expertise have triggered a rapidly booming global job engine.

The political targets for 2020 have been no less ambitious than the slogans of the globally active companies: voRWEg gehen is a significant example of this.

The fact that the global energy suppliers and also the regional, interconnected energy suppliers are communicating the current situation of the energy transition with more than unhappiness is justified on closer inspection.
A company in the private sector would neither be able nor want to continue a production line if it could not even break even.

So why are energy suppliers expected to operate their power plants on the grid if feeding renewable energy into the grid must take priority if the energy transition is to be a success?

Because security of supply and the uninterrupted provision of energy are an indispensable pillar, both for industry and for private households. However, energy supply companies are also subject to the demands of global competition on the markets and cannot survive without economic success.

The slogans of companies such as:
+ "Power for new things" (Evonik),
+ "voRWEg gehen" (RWE),
+ "New Energy" (E.ON)
have been both a program and a claim with the start of the energy turnaround in Germany.

However, the competencies of RWE, from 2016 also with Innogy, and the competencies of E.ON have not complemented each other; rather, RWE and E.ON have been in competition with each other in the past. This has changed.

With their development, RWE and E.ON have taken an important step towards creating these sustainable structures.

It was only in 2018, after RWE and E.ON had agreed on a very comprehensive, substantive coordination of their business objectives, that the changeover created a basis for the two companies to complement each other:

+	RWE had fully transferred its Innogy shares to E.ON by September 2019,
+	RWE had received the entire renewable energy business from both E.ON and Innogy in return,
+	so that RWE pursues energy generation including renewables,
+	so that E.ON operates the grids.

Power-to-gas and methanation are not only on the list of potentials for globally operating energy supply companies such as RWE and E.ON in order to maintain the security of supply of electricity, gas and heat in the future.

Energy storage systems are undisputedly a necessary component of a sustainable energy infrastructure, both for the excess electrical capacity generated from renewable sources and for energy transportation, as well as for the provision of energy on site. Short-term storage and long-term storage complement each other and enable energy supply security in a sustainable energy infrastructure.

Virtual power plants, smart meters, mini-grids and power-to-X are no longer alien concepts.

Wind turbines, wind power electrolysis, solar energy and heat-generating solar modules have become increasingly important in recent years.

In the past, energy supply companies have struggled to make concrete statements about the energy transition.
In the past, the requests to the energy supply companies for concrete impulses in the area of the current target agreements had led to a constructive dialog with RWE and also with E.ON.
The energy supply companies were happy to provide support in the potential grid of the energy transition if the municipalities and regions also supported the projects.

It has been no secret that the success of the energy transition will also be linked to energy efficiency. The numerous developments in energy systems also show the extent to which energy efficiency has increasingly influenced our everyday lives in the age of digitalization with applications for heat, electricity and gas.
The clever use of intelligent information and communication technologies has become important for modern energy supply and for achieving a sustainable energy infrastructure.

It is not only the younger generations who are intensively involved in the energy transition.

This is also reflected in the questions asked by people of all ages, for example during the trade fairs:

+ driven by curiosity,
+ through to career prospects,
+ questions about the range of current energy sources,
+ to the expected cost trends for energy,

the questions are characterized by uncertainty and great interest at the same time.

Energy storage systems are the indispensable and supporting components of a sustainable energy infrastructure.

The storage of electricity is a key component for the success of the energy transition in the long term. The increasing expansion of renewable energy generation would place an even greater burden on the electricity grids; against this backdrop, fluctuating energy generation will be even more pronounced on windy days in summer and on windless days in winter.

The short-term solution:

Electricity storage systems are increasingly complementing conventional energy generation through smart grids.

<u>The long-term solution:</u>

The electricity storage systems are networked with each other in meshes, which together make up the virtual power plant.
Both the global challenges and the local challenges together are the driving force behind the energy transition. The year 2022, with Russia's invasion of Ukraine and the start of the resulting crisis, is the trigger for the already formulated goals of the energy transition to be pursued even more ambitiously.

The publications and analyses relating to hydrogen as an energy source show that the German government's greenhouse gas reduction target for 2030 of 55% compared to 1990 is achievable with the use of hydrogen-based energy concepts.

The profitability analyses clearly show that the comparative cost level in the fuel sector also favors the use of hydrogen in the mobile sector.
However, in order to achieve the targets set for reducing greenhouse gases and switching to renewable energies by 2050, considerable investment in infrastructure is required, whereby the expected follow-up costs of increasing global warming will be significantly higher.

The generally recognized drivers of changing energy technologies worldwide are climate change, local emissions, security of energy supply and industrial competitiveness; the relevance varies depending on the country in question.

Following the nuclear power plant accident in Fukushima, which was triggered by a natural disaster, several countries have turned away from nuclear power. One of the main reasons for this was certainly the realization that if a world leader in nuclear power such as Japan cannot prevent an accident, other countries also face an incalculable risk from nuclear power generation.
In addition, there is the major problem of the safe storage of spent fuel rods from nuclear power plants, as well as the enormous perpetual costs of the storage facilities.

In Germany, this event led to a broad political consensus among all parties against the continued use of nuclear power. At the same time, however, greenhouse gas emissions are to be greatly reduced.
The German government's declared commitment to reduce greenhouse gas emissions by 80% to 95% by 2050 compared to the reference year 1990 is a very ambitious target and may not be fully achievable given the timing. However, the globally booming job engine has already been created; also with German technology know-how and German infrastructure know-how.

The reduction target for 2030 is 55% compared to the reference year 1990. This means that the energy industry and energy technologies are facing global, and subsequently local, challenges that are also linked to the targeted reduction of environmental impacts and economic policy objectives.

With the economic policy objectives, the German government is promoting renewable energies, electromobility, combined heat and power generation and sustainable energy technologies.

The path of the energy transition is not only the path away from material-based, fossil fuels, bound up in coal, oil and natural gas, towards renewably generated energies, the path of the energy transition is also the path away from grid centrality with power plants towards interconnected, decentralized energy generation.

The bridging technologies are being optimized at the technical tangents, which certainly includes the development of more efficient, central power plants, which certainly includes the optimization of the gas infrastructure.

+ The technologies of the future are undergoing system integration at the technical tangents of the energy transition,
+ Digitalization links decentralized energy generation with the aim of "virtual power plants",
+ the gas infrastructure transports generated product gases and supports, for example, power-to-gas and methanization.

With reference to the German government's stated demand for reduced greenhouse gas emissions, renewable energies and electromobility play a prominent role in achieving the ambitious targets.

In this context, the further expansion of renewable energies is being very closely monitored by the EU, the federal ministries and the state ministries, whereby the storage of electrical, renewable energy with hydrogen is becoming increasingly important due to the fluctuating feed-in of wind and solar power. The focus is always on ensuring that the electrical energy generated from renewable sources is always used directly and that the excess capacity, the electrical energy that the electricity grids cannot absorb, is stored with hydrogen for times when less electrical energy can be generated from renewable sources. As a result, the aim is to use hydrogen directly.

In fact, the industrial sectors have already developed utilization paths for hydrogen:

+   the storage of excess capacity generated from
       renewable sources in the electricity market,
+   as a fuel for vehicles with highly efficient fuel cell
       drives,
+   feeding hydrogen into existing natural gas grids with
       power-to-gas,
+   the development of new energy products with
       methanation, for example.

The hydrogen utilization paths investigated show that the use of hydrogen in fuel cell vehicles offers the greatest reduction in CO2 emissions, as the change in technology means that the fuel is used at a higher level of efficiency compared to the combustion processes in petrol and diesel engines on the one hand, and a fuel with significantly higher CO2 emissions is avoided in mobile applications compared to natural gas combustion on the other.

A decisive factor for the success of the development of future-proof drive concepts is that the comparison with the performance characteristics of today's vehicles generates recognizable advantages.

The criteria for the acceptance of future-proof mobile applications are:
+ the range,
+ the refueling time,
+ the dynamics of the drive
+ technical reliability,
+ the purchase costs,
+ the operating costs,
+ the recuperation.

Recuperative braking is new in the overall view of mobility; only with electromobility is energy recovery through the braking process possible, both with battery vehicles and with fuel cell vehicles.

A frequently cited goal of the energy transition is that:
+   wind energy with approx. 80%,
+   other renewable energies with approx. 10%,
+   gas products also contribute approx. 10%
        to total electricity generation in the EU, whereby:
+   approx. 65% of electricity generated from renewable
        sources is used directly,
+   approx. 35% of the electricity generated from
        renewable sources supports hydrogen production.
        The amount of hydrogen produced would be
        sufficient, for example, to power around 28 million
        cars.

Based on the mutually confirming calculation results from
the research work and the development work in the EU,
an economic analysis of the energy concept with the use
of hydrogen as an energy carrier is possible.

With a view to the expected costs for hydrogen, the
question of how the hydrogen is transported from the
production site to the consumers as an energy storage
medium is also crucial.

Hydrogen can be transported in both liquid and gaseous
form. Technically mature is the transportation of hydrogen
in pipelines or by truck; this is the current, industrially
successful practice. In the long term, transporting gaseous
hydrogen via pipelines is the most economical way of
introducing large quantities of hydrogen into the transport
sector.

Suggestions

With the books I have published, currently 26 books, you will find supplementary and very detailed information in the areas of energy, climate and the environment.
I publish the books with BoD Verlag and all books are available from all book publishers and also from online marketplaces such as Amazon.

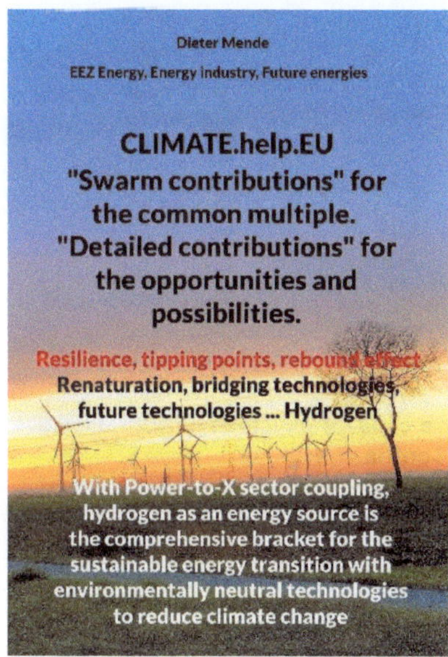

Books:
Left: 76 Book-Pages A4, CLIMATE.help.EU.
Right: 68 Book-Pages A4, ENVIRONMENT.help.EU.

Book:
Left: 88 Book-Pages A5, The energy turnaround: perhaps seeming paradoxical at first? Current positions.
Right: 60 Book-Pages A5, The job engine oft he energy transition. The global starting position.
Below: 116 Book-Pages A5, Patient Europe: a case for psychiatry? No! Release the handbrake, the hedgehog posture afer Corona.

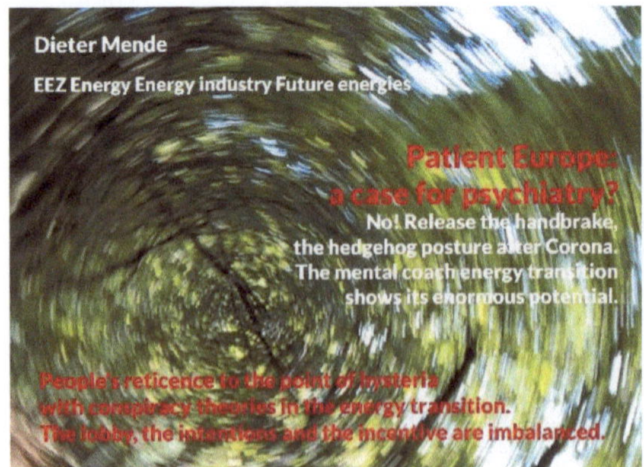

The author

Dieter Mende

Homepage:
www.eez-mende.de

The book is written with the background of my professional career in chemistry as well as in electrical engineering; supplemented with the background of the energy dialog EEZ Energie Energiewirtschaft Zukunftsenergien, which I founded myself on 05.07.1995.

The professional basis is broadly diversified. In both cases, the employers noticed the very successful energy dialog EEZ Energie Energiewirtschaft Zukunftsenergien, so that the jobs arose from it.

Since September 1997 in the project office of a regional energy supply company, responsible for the central control technology and automation technology for the energy centers. Changed departments in November 2019 with the project planning of electrical power grid construction.

Since February 2003, initially commissioned with the tasks of setting up and expanding the regional hydrogen nucleus h2herten, then commissioned with the tasks of project acquisition and company acquisition, networking, market communication, in the team of the hydrogen user center h2herten.

What drives me to write reports and books is my passion for highlighting the opportunities and possibilities in the potential grid of the energy transition, with hydrogen as an energy carrier; my ambition to build and expand a hydrogen infrastructure with the advertising of cross-industry service providers, with the identification of sustainable contributions and the resulting complementary competencies.

I met the established companies in the energy market and their initial rejection of hydrogen as an energy source and the fuel cell as an energy converter with meaningful results of potential analyses, with the potential of Power-to-X sector coupling, the identification of unique selling points and with the complex project initiatives of the multi-layered interests of all those involved.

With perseverance and patience, even initial skeptics of hydrogen as an energy source and the fuel cell as an energy converter were successfully won over to a complementary and sustainable infrastructure in the energy transition. I successfully contributed to the acquisition of members for the foundation of an advisory board for h2herten, from which the advisory board for the h2-netzwerk-ruhr emerged through expansion in 2008.